Just Call me HONEY

Written By: Linda Smock
Pictures by: Jennifer Singer

buzzing with life

THIS BOOK BELONGS TO:

_____

·······································

_____

Your Name

TABLE OF CONTENTS

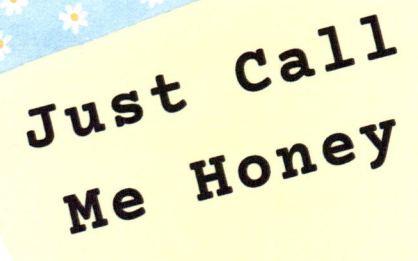

Just Call Me Honey

Hi, my name is Busy Bee. You can just call me "Honey." That's what my momma does. She's the queen.

6

She tells me to act like other bees. She says that means to work hard. Our work is essential. That means our work is needed for people to live.

May I tell you about my family?
May I tell you about my work?
I hope you will read my story.

8

First, I want you to know about all my family. I'm in the bee family. There are over 20,000 species of bees. A species is a group of things with the same name. Our name is "bee." But remember, you can call me Honey. That's what my momma does. She's the queen.

9

Eight of the species are honey bees. There were eleven. Three have not been seen for many years. They are extinct. That means they have all died away. That is because you humans need space. You use it for houses and fields. The bees then don't have enough food. The wildflowers are not there.

The bees can't eat. They starve to death.

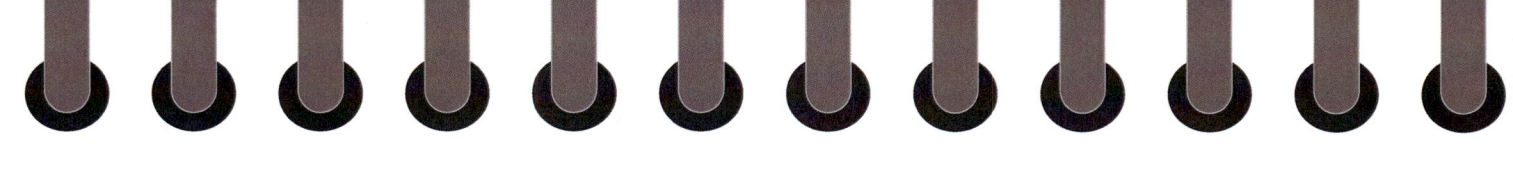

The eight species of honey bees all make honey. My momma, siblings, and I are western honey bees. We are domesticated. That means people keep us. We honey bees are not pets. Humans count on us to pollinate plants. They like our honey. They use the beeswax we make.

Let me tell you about my momma. She's a single mom. She lives longer than her children. She can live for up to five years.

Her main job is to lay eggs. First, she flies out looking for a drone (boy bee). Drones fertilize the eggs inside her. Then she lays the eggs. Some eggs become girls. Some become boys.

A few girls are selected
to be future queens.
They are fed a special
diet. The new queens will
start their own colonies.

My Home

We honey bees can make many places our home. We like those hives that you humans make. You may see them from the highway when you take a road trip. They are lined along the edge of a field.

We can make our home in a hollow tree.

Sometimes, we find holes in concrete blocks stacked together. We use all those holes to build honeycombs. You might see us in a concrete wall.

Occasionally,

we find an empty birdhouse and use it.

20

Sometimes, you see a big ball of us on a tree limb. That means we are moving. We are searching for a new place to live. When we find it, we will move in. You may never see us again.

MY LIFE CYCLE

My momma, the queen, lays eggs. They hatch. They become larvae. Larvae look like little worms. Their big sisters feed them a little bit of royal jelly. (The big sisters make the royal jelly. They have special glands to make it for the babies.) The larvae soon develop into bees. Some are boys, or drones. Most of the larvae become girl bees. A few are fed a special diet of only royal jelly. These become queens.

The rest of the girl bees are called worker bees. Worker bees start their lives as hive bees. Sometimes, hive bees are called nurse bees. Then, they become house bees. Last, they become field bees, or forager bees. I'll tell you about each stage.

Worker Bees

Hive or Nurse Bees

House Bees

Forager Bees

When the larvae hatch, they are not strong enough to fly from the hive. This is when they are called hive bees or nurse bees. They clean the central part of the hive. They feed the babies. They feed the drones. They feed the queen. Queen Momma is strict. The house must be kept clean. The hive bees do all that.

Next, the girls move to the outer section of the hive. They still don't fly. They are called house bees. They build honeycombs. They receive nectar from foraging bees like me. They take the nectar I gather. They mix it with a liquid they produce to turn it into honey. They store the honey. They make sure the hive gets fresh air. Occasionally, a bee dies. They must be taken out of the hive. The house bees do that.

Finally, they are ready to become forager bees. They look for pollen and nectar. I was doing that before I stopped to tell you this story. Forager bees get up when the sun comes up. They fly a mile or two from their home. They find a patch of flowers. They start to gather the nectar. While they are gathering nectar, they pollinate the flowers. ("Pollinate" means to put pollen on the flowers so they can produce seeds.) The foragers fly the nectar back to the hive. They give it to a younger bee, a house bee. They fly back to the flowers. They continue this all day long.

29

Do you know why I fly far from my hive to forage in flowers? So I don't give away the location of Momma.

She is the queen. We must protect her. The future of the colony depends on her.

My brothers, the drones, have a much easier life. We sisters feed them and take care of them. Their only job is to go looking for a queen. When they find her, they help her make new babies. They don't have the fun of foraging. If food is short, the boys leave the hive. We must have food for the queen and the babies.

By the way, if you see a drone or boy bee, you may think he is a fly. His eyes bulge out. He does not look like me. He does not have a stinger either.

HOW MY SISTERS AND I MAKE HONEY

I work hard to make honey. That's why I'm called Busy Bee. (Remember, you can call me Honey. That's what my momma does. She's the queen.)

34

I buzz from flower to flower. I like big yards and fields with one kind of flower. I gather the pollen and the nectar. When I land on the flower, the pollen gets on my legs and hair. (Yes, I have tiny, short hairs on part of my body.) I fly to another flower and land. The pollen from the flower I was just on gets on the new flower. That is called pollination. The flower can then become a seed or a fruit.

Can you name some plants that need to be pollinated?

Think about nuts, vegetables, and fruits. Think about the grass that a cow eats. Think about pretty flowers.

37

I suck the nectar from the flower to make honey.
Nectar is a rich sugar liquid. I use my proboscis to
get the nectar. My proboscis is like a tongue.

Then I swallow the nectar to my crop. My crop is similar to a stomach. As the nectar goes down, I put the right things into it to make it honey.

When I return to the hive, I transfer the honey to a house bee. She adds more liquid to it. It goes to her crop.

Later, she returns it and gives it to another house bee.

After several bees do this, they add a little water. Then, it is honey. It is stored in the beehive. The house bees seal it with beeswax. It will last for thousands of years.

I fly several miles a day. I get hungry. It takes energy to fly. I also can carry pollen that weighs about 35% of my body weight. That takes a lot of energy.

Do you know where I get all that energy? From honey, of course. That is why God created us to make honey. We need many calories. We are busy bees.

44

Let's stop and think for a minute. If you weigh 50 pounds, what is 35% of your weight? Yes, that is 17.5 pounds. That is more than two gallons of milk. Could you walk and carry two gallons of milk two miles? You would have to be very strong.

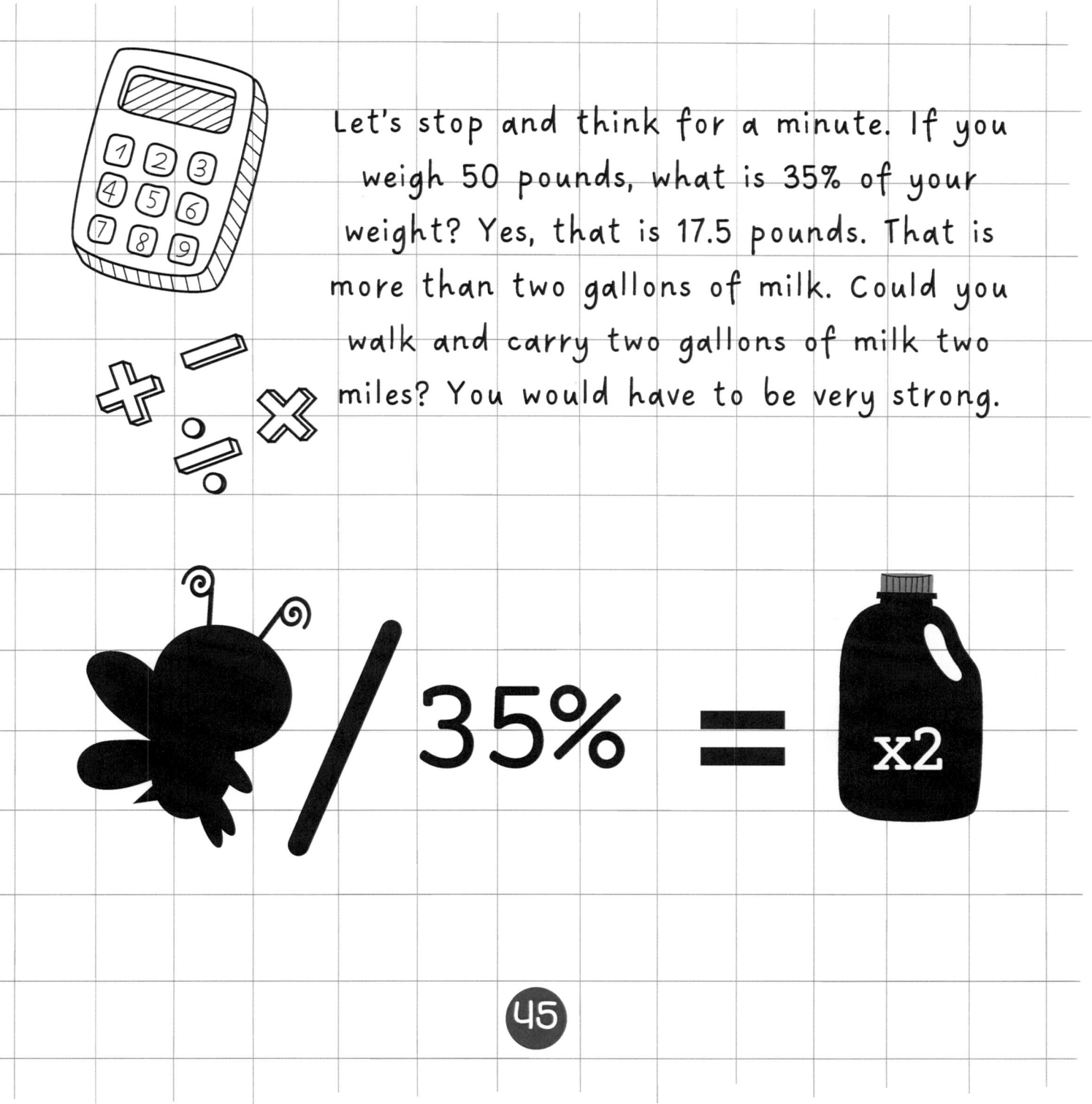

35% = x2

As a forager bee, I protect my momma's hive. We must always protect her.
(She's the queen.)

47

To protect the hive, sometimes bees have to sting something. Did you ever read about a bear being stung by bees? What is his name? He got stung because he robbed the bees of their honey!

Honey helps bears store up fat for the winter. Bees like me sting their noses when they try to rob us. They don't like that.

Sometimes, you humans get too close to our hive. I buzz around you. If you don't move away, I will sting you. If you hit me,

I belong to a colony that is domesticated.
Do you remember what that means?
Humans have to move us. Sometimes, they
check our hives for mites and beetles.

People also need to gather honey. They need to do something to keep us from stinging them. Smoke confuses us, so people use it. We can't talk to each other when there is smoke. We don't overreact. We don't get upset and sting you. You can gather the honey. You can check for pests like hive beetles. You are safe. We are safe.

HOW I MAKE BEESWAX

Worker bees have eight glands that make beeswax. It is pushed out of the abdomen of the worker bees. It is in the form of scales.

The worker bees collect it. They form the beeswax into hexagon cells. We forager bees cannot make beeswax. Those eight glands no longer work in us. The worker bees are good at making beeswax.

The worker bees know the difference between a circle and a hexagon. A hexagon has six sides.

Would you draw a circle and a hexagon for me?

Trace and Draw

Directions: Trace the shape
and then draw the shape.

You Trace

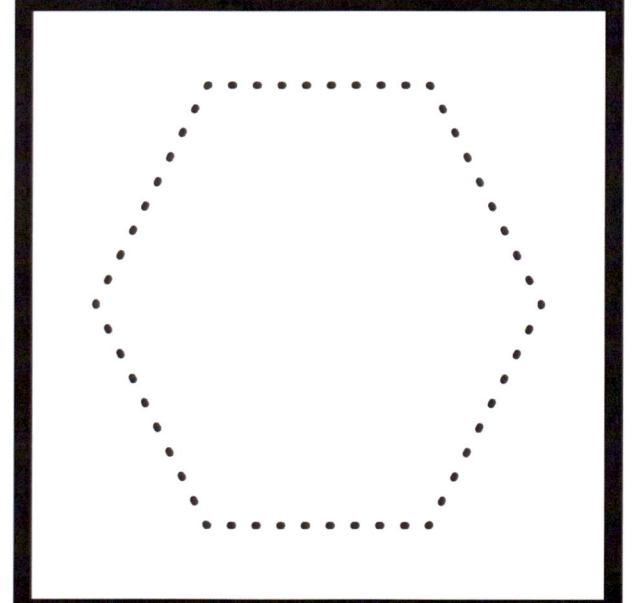

You Draw

Trace and Draw

Directions: Trace the shape
and then draw the shape.

You Trace

You Draw

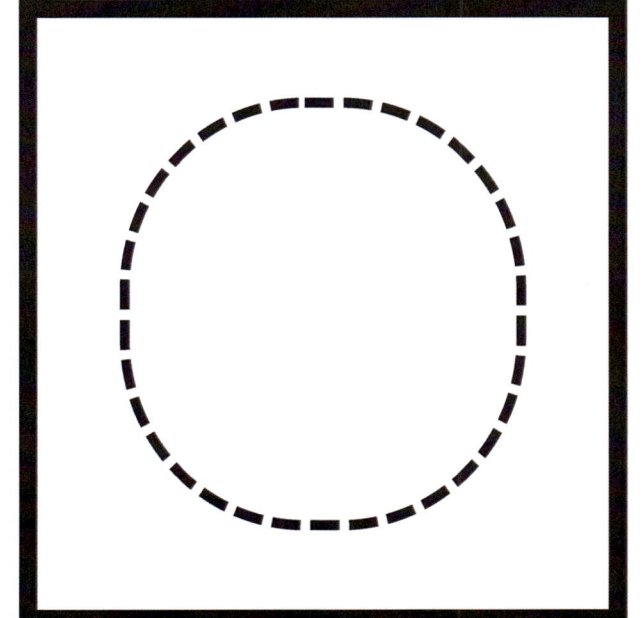

The worker bees secrete the wax. They warm it with their bodies. They first form it into a circle. Then, they reshape it into a hexagon. They use the heat of their little bodies to add their hexagon to other hexagons. They work together to make a honeycomb. Other worker bees bring honey. It is stored in the hexagon spaces in the honeycomb. Worker bees will seal each little hexagon. Beeswax can keep the honey safe for thousands of years.

Well, so long for now. I'll see you in the flowers. Remember, my name is Busy Bee. You can just call me Honey. That's what my momma does. She's the queen.

THE END